Three Firefighters Die in
Pittsburgh House Fire
Pittsburgh, Pennsylvania

Investigated by: J. Gordon Routley

This is Report 078 of the Major Fires Investigation Project conducted by Varley-Campbell and Associates, Inc./TriData Corporation under contract EMW-94-C-4423 to the United States Fire Administration, Federal Emergency Management Agency.

Homeland Security

Department of Homeland Security
United States Fire Administration
National Fire Data Center

U.S. Fire Administration Fire Investigations Program

The U.S. Fire Administration develops reports on selected major fires throughout the country. The fires usually involve multiple deaths or a large loss of property. But the primary criterion for deciding to do a report is whether it will result in significant "lessons learned." In some cases these lessons bring to light new knowledge about fire--the effect of building construction or contents, human behavior in fire, etc. In other cases, the lessons are not new but are serious enough to highlight once again, with yet another fire tragedy report. In some cases, special reports are developed to discuss events, drills, or new technologies which are of interest to the fire service.

The reports are sent to fire magazines and are distributed at National and Regional fire meetings. The International Association of Fire Chiefs assists the USFA in disseminating the findings throughout the fire service. On a continuing basis the reports are available on request from the USFA; announcements of their availability are published widely in fire journals and newsletters.

This body of work provides detailed information on the nature of the fire problem for policymakers who must decide on allocations of resources between fire and other pressing problems, and within the fire service to improve codes and code enforcement, training, public fire education, building technology, and other related areas.

The Fire Administration, which has no regulatory authority, sends an experienced fire investigator into a community after a major incident only after having conferred with the local fire authorities to insure that the assistance and presence of the USFA would be supportive and would in no way interfere with any review of the incident they are themselves conducting. The intent is not to arrive during the event or even immediately after, but rather after the dust settles, so that a complete and objective review of all the important aspects of the incident can be made. Local authorities review the USFA's report while it is in draft. The USFA investigator or team is available to local authorities should they wish to request technical assistance for their own investigation.

The fire investigation reports are developed by USFA staff and by Varley-Campbell & Associates and TriData Corporation, their staff and consultants, who are under contract to assist the Fire Administration in carrying out the Fire Reports Program.

Appreciation goes to Chief Charlie Dickinson and all of the members of the Pittsburgh Fire Bureau for their assistance in preparing this report. The author wishes to acknowledge the assistance that was provided by the members of the Board of Inquiry, headed by Deputy Chief John Gourley, and the support of the officers and members of Local 1, International Association of Fire Fighters.

For additional information contact Mr. Kenneth Kuntz, Major Fires Investigations Project Officer at the U.S. Fire Administration, 301-447-1000.

For additional copies of this report write to the U.S. Fire Administration, 16825 South Seton Avenue, Emmitsburg, Maryland 21727. The report is available on the Administration's Web site at http://www.usfa.dhs.gov/

U.S. Fire Administration

Mission Statement

As an entity of the Department of Homeland Security, the mission of the USFA is to reduce life and economic losses due to fire and related emergencies, through leadership, advocacy, coordination, and support. We serve the Nation independently, in coordination with other Federal agencies, and in partnership with fire protection and emergency service communities. With a commitment to excellence, we provide public education, training, technology, and data initiatives.

TABLE OF CONTENTS

Three Firefighters Die in
Pittsburgh House Fire
February 14, 1995

Local Contact: Chief Charlie Dickinson
                   Pittsburgh Bureau of Fire
                   200 Ross Street, 5th Floor
                   Pittsburgh, PA. 15219
                   (412) 255-2860

OVERVIEW

Three Pittsburgh firefighters died on February 14, 1995, when they ran out of air and were unable to escape from the interior of a burning dwelling. The three victims were all assigned to Engine Company 17 and had advanced the first hoseline into the house to attack an arson fire in the basement. When found, all three were together in one room and had exhausted their air supplies. Three other firefighters had been rescued from the same room, which caused confusion over the status of the initial attack team.

This incident illustrates the need for effective incident management, communications, and personnel accountability systems, even at seemingly routine incidents. It also reinforces the need for regular maintenance and inspection of self-contained breathing apparatus, emphasizes the need for PASS devices to be used at every fire, and identifies the need for training to address firefighter survival in unanticipated emergency situations.

This incident also reinforces a concern that has been identified in several firefighter fatality incidents that have occurred where there is exterior access to different levels form different sides of a structure. These structures are often difficult to "size-up" from the exterior and there is often confusion about the levels where interior companies are operating and where the fire is located. In these situations it is particularly important to determine how many levels are above and below each point of entry and to ensure that the fire is not burning below unsuspecting companies.

1

SUMMARY OF KEY ISSUES

Issues	Comments
Incident Command	Command was not established by the first arriving company. The acting battalion chief was coming from another call and had a delayed arrival. All first alarm companies had self-committed before the acting battalion chief assumed command of the incident.
Accountability	Accountability procedures were not implemented. The locations and functions of companies operating inside the house were not known to the Incident Commander. It was not realized that members were missing.
Crew Integrity	All crews did not function as single tactical units. Some of the individual members from these companies performed unrelated tasks and were not under the supervision of their company officers. Most of the personnel were working in temporary assignments for that shift.
Emergency Survival	The actions of the three victims when they realized they
Action	were in trouble are not known; however, they do not appear to have initiated emergency procedures that could have improved their chances of survival or made other firefighters aware of their need to be rescued.
Rapid Intervention	Some fire departments have adopted procedures to assign a Teams rapid intervention team at working fires. The objective of this team is to be ready to provide immediate assistance to firefighters in trouble.
Communication	There was a lack of effective fireground communications at this incident. There was no exchange of information with the interior crews after they entered the dwelling. All of the first alarm companies were operating before the acting battalion chief arrived and assumed command. The Incident Commander did not receive any progress reports from these companies.
Portable Radios	Although two of the three firefighters who died had portable radios, they do not appear to have attempted to use them to summon assistance. (One of the radios was found to be inoperative after the incident.)
Interagency Coordination	The communications problems were complicated by the fact that EMS units on the scene of a fire report to their own supervisors and communicate on their own radio channels. The interaction and communications between fire and EMS units were inadequate.
SCBA Maintenance	Examination of the SCBA units used by the three victims and one of the injured firefighters indicate that improvements are needed in maintenance, inspection, and testing programs.
PASS Devices	All three victims had PASS devices, however they were not turned on. It is likely that a functioning PASS device would have alerted other firefighters in the immediate area to the unconscious victims.

BUILDING DESCRIPTION

The Homewood section of Pittsburgh is a densely populated residential area. Most of the structures in the area are one- and two-family wood frame dwellings that are at least 75 years old. The streets are narrow and the houses are closely spaced. Most of the houses appear to be two stories, but many are built on sloping ground and have additional floor levels that cannot be seen from the street.

The fire occurred in a single family dwelling that was approximately 20 feet wide and 33 feet deep and had three occupied floor levels above a partially finished basement. The interior walls were lath and plaster, and the exterior walls were aluminum siding over asphalt shingles attached to balloon

EAST SIDE OF DWELLING - 8361 Bricelyn Street

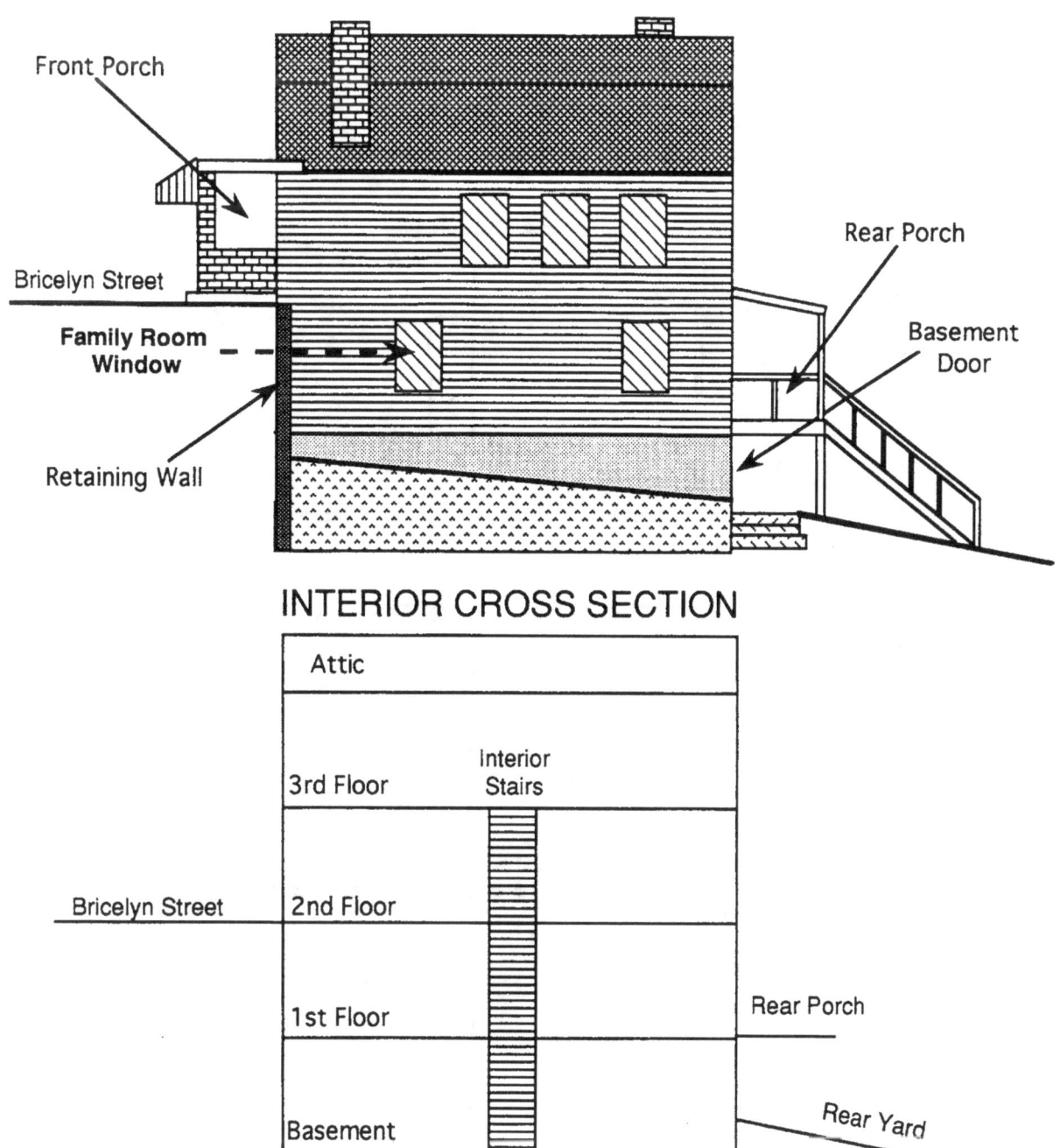

Front Porch

Bricelyn Street

Family Room Window

Retaining Wall

Rear Porch

Basement Door

INTERIOR CROSS SECTION

Attic		
3rd Floor	Interior Stairs	
2nd Floor		
1st Floor		Rear Porch
Basement		Rear Yard

Bricelyn Street

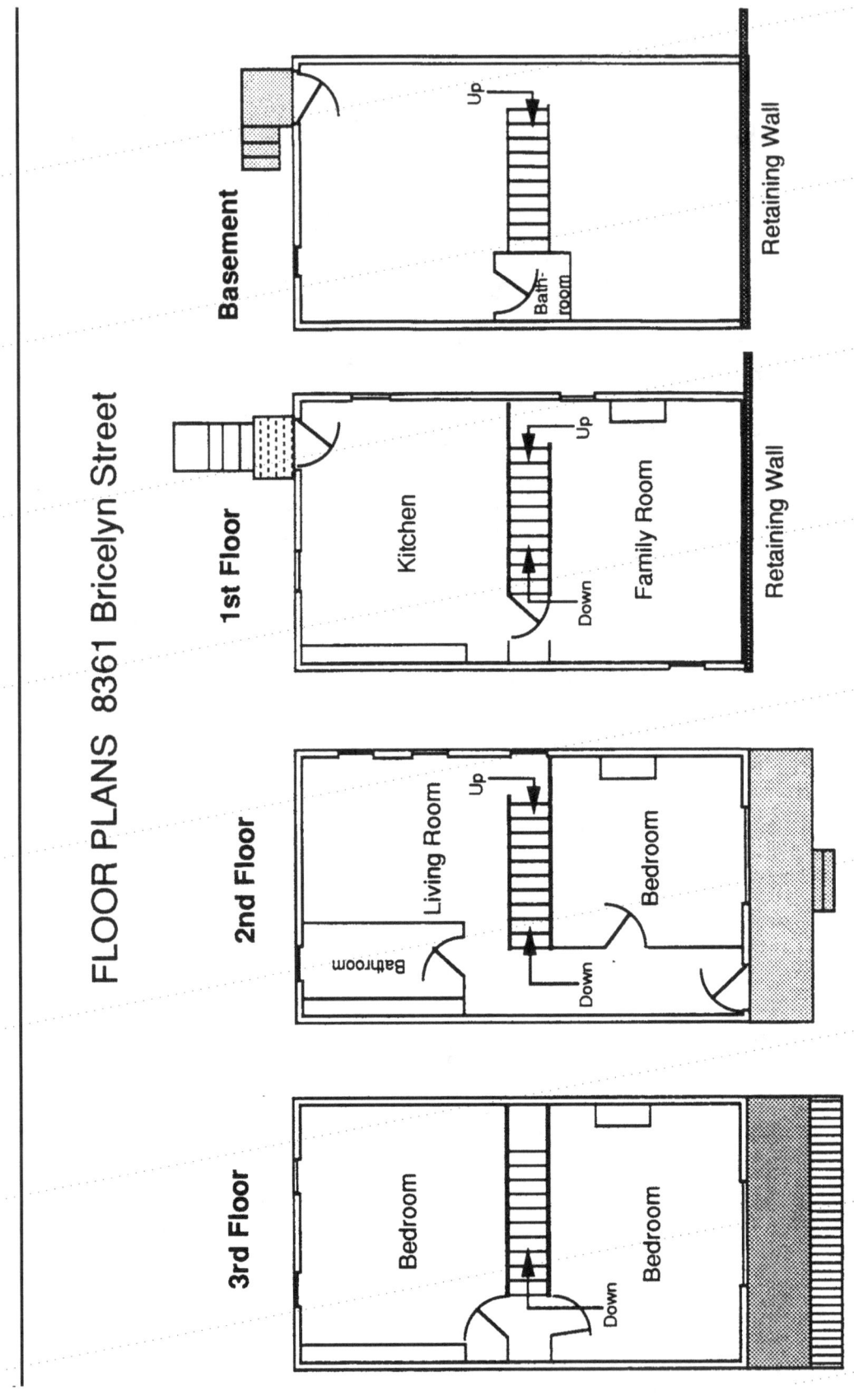

FLOOR PLANS 8361 Bricelyn Street

Basement

1st Floor

2nd Floor

3rd Floor

frame wood construction. The separation to the neighboring house on the left side was only 27 inches wide, which created a significant exposure problem. On the right side there was a 12 foot wide side yard between houses.

Because the house was built on a slope, the front entrance at 8361 Bricelyn Street was actually at the second floor level. When viewed from the front, the house appeared to have only two stores; however, from the rear most of the basement was exposed above grade level and the house appeared to be the equivalent of four stories. The first floor was one level up from the back yard and one level down from the front door. (Refer to the building diagram and floor plans on the following pages for details of the arrangement.)

A stone retaining wall was built along the front of the property and served as the front wall of the basement and first floor levels. The side yard was approximately 12 feet lower than the sidewalk in front of the house.

Doors to the exterior were provided on three levels. The front door, on the second floor, opened to a narrow front porch adjacent to the sidewalk. There was a rear porch with exterior stairs from the first floor down to the back yard. Immediately below the rear porch was an access door into the basement.

The kitchen and family room were located on the first floor, the living room and one bedroom were on the second floor, and two additional bedrooms were on the third floor. The heating equipment, a washer and dryer, and storage areas were located in the basement. The furnace, water heater and clothes dryer were all fueled by natural gas.

The three stair sections were stacked, one directly above another, and divided each floor into front and rear sections. A wood door at the top of the basement stairs opened into a narrow vestibule on the first floor; when this door was open it blocked the opening between the kitchen at the rear and the family room at the front. There were no doors in the stairways between the first and second or the second and third floors.

The basement had one small window facing the rear yard. On the first floor, there were two windows in the family room, one on each side of the house, and two in the kitchen. The family room windows were approximately five feet above grade level and had been covered-over on the inside with a clear acrylic plastic (plexiglass) 3/32 inches thick to create the effect of storm windows. There were also windows on the second and third floors.

The house was occupied by two adults, one teenage, and one child; only one adult and the child were at home when the fire was discovered. The contents were ordinary household furnishings. The rooms were small and crowded with furniture. The home was equipped with battery powered smoke detectors.

THE FIRE

The weather on the night of February 14 was cloudy with a moderate wind and a temperature of approximately 15 degrees Fahrenheit. The ground was partially covered with snow at the time of the fire.

Fire Origin and Spread

The fire was determined to have been caused by arson and originated in the basement, directly under the kitchen. This would place the point of origin close to the gas-fired appliances. The fire first involved contents in the basement, then extended upward through void spaces in the exterior walls and around the stairway. The void spaces allowed the fire to extend directly to the attic. It later broke through the walls and/or ceiling and involved the bedrooms on the third floor.

There was some fire involvement in the kitchen, particularly under the counters, which were attached to the wall where the largest void space was located. A partial floor collapse occurred in the kitchen directly over the point of origin, during the later stages of the fire. The fire extended via the walls from the basement directly to the attic, then involved the third floor; there was no fire involvement in the family room or in any of the rooms on the second floor.

The occupants reported that the first indication of fire was smoke seeping into a bedroom on the upper floor. The occupants exited from the house and called 9-1-1 from a neighbor's telephone to report the fire. The first alarm assignment was dispatched to 8366 Bricelyn Street at 0022 hours.

Response

The initial response consisted of Engines 17, 18, and 19, Truck 17, and Battalion Chief 2 (Unit 4042). Engine 19 was dispatched in place of Engine 15, which was on another call, and Battalion Chief 2 was responding in place of Battalion Chief 3. Battalion Chief 3 (Unit 4043) became available after the call was dispatched and replaced Battalion Chief 2 on the assignment. The Emergency Medical Services Division dispatched Medic 11 and Rescue 1 on the first alarm, each with two paramedics on board. (Pittsburgh EMS is a separate division of the Department of Public Safety.)

The first alarm assignment included a total of 17 personnel, including the acting battalion chief. The three engine companies and the truck company each had four personnel on duty. On this particular night, only fire of the 17 Fire Bureau personnel responding on the first alarm were in a normally assigned position; all of the others were temporarily detailed, acting in a higher classification, working on overtime, or had traded shifts or assignments with another member for the night.

Engine 17 was staffed by a captain and three firefighters. The captain and the driver were in their regular assignments, while the other two firefighters had been detailed from Engine Company 8. The officer in charge of Truck 17 was a firefighter moved-up to acting lieutenant, while a captain who is normally assigned to Engine 8 on a different shift was working on overtime and riding as a firefighter on Truck 17. He had previously been assigned as a captain on Engine 17.

The crews of Engine 18 and 19 also included several personnel who usually rode in different positions. A battalion chief was acting as the deputy chief in charge of the shift and a captain was acting as Battalion Chief 3.

Arrival

Engine 17 was first to arrive at 0027 hours, just ahead of Truck 17. The homeowner was standing outside and directed Engine 17 to 8361 Bricelyn Street. The captain went to the front porch to investigate and heavy smoke became evident when the front door was opened. Truck 17 entered from the opposite end of the block because Bricelyn Street is very narrow and provides room for only a single traffic lane between parked cars. (See fire scene diagram on following page.)

The captain entered the house briefly, then returned to the front porch and motioned to his crew to advance a 1-3/4 inch preconnect line. He used his portable radio to report that Engine 17 was on the scene with "smoke from the first floor" and directed Engine 15 to lay a supply line to Engine 17.

Recognizing that Engine 15 was not on the assignment, the dispatcher relayed this assignment to Engine 18. The dispatcher did not advise Engine 17 that Engine 15 was not responding on the call. In this area Engine 15 usually arrives very soon after Engine 17, while Engines 18 and 19 take several minutes longer to reach the location.

The captain and two firefighters from Engine 17 went in the front door with the initial attack line, all wearing full protective clothing and self-contained breathing apparatus, while the pump operator remained outside. The 1-3/4 inch hoseline was advanced to the top of the stairs and then made a right turn to go down the stairs to the family room. It is not known if the crew spent any time on the second floor before they went down to the first floor.

Truck 17 began operations by laddering the front of the structure with portable ground ladders and ventilating the third floor windows above the front porch. The aerial ladder could not be used because of overhead power lines. Two of the crew members then used a ground ladder to go to the roof to begin vertical ventilation. Another crew member from Truck 17 worked with two crew members from Engine 18 to lower a ladder down to the side yard of the house to provide access to the rear yard.

Engine 18 had arrived at 0032 hours. The crew stretched a 5 inch supply line from a hydrant and connected it into their pump, and then stretched another 5 inch line from Engine 18 to Engine 17. After positioning the ground ladder, they advanced a 1-3/4

inch attack line from their apparatus down the ladder to the side yard and around to the rear of the involved house. They observed the fire in the basement and operated the hose- line through the rear basement window. The captain went to the roof to assist Truck 17, leaving the two firefighters to operate the line.

Engine 19 reported on the scene immediately after Engine 18 and parked behind Engine 18. The crew advanced a second 1-3/4 inch attack line from Engine 17 to the front porch. At the front door they encountered heavy smoke and heat which prevented them from advancing.

The acting battalion chief arrived at 0037 hours and took a position at the front of the house. At 0038 hours he reported "...a three story house, exposures 2 and 4,...right now we have heavy smoke coming out of the second and first floor."

At 0039 hours Medic 11 reported on the scene and assumed a stand-by position near the front of the house. Rescue 1 arrived several minutes later and was directed by Medic 11 to take a position in the rear alley (Exley Way).

Fire Extension

At 0046 hours the acting battalion chief requested a second alarm, based on his observation of heavy smoke coming from the third floor and the attic. He directed the second alarm companies to come in via the alley in the rear.

The crew of Truck 17 was on the roof and could see the increasing smoke and heat conditions in the attack. The captain who was working as a firefighter on Truck 17 advised the acting lieutenant that the fire was in the walls and was coming up through the void spaces. He said that he would go inside to check on fire conditions on the lower floors and advise Engine 17 of the vertical extensions.

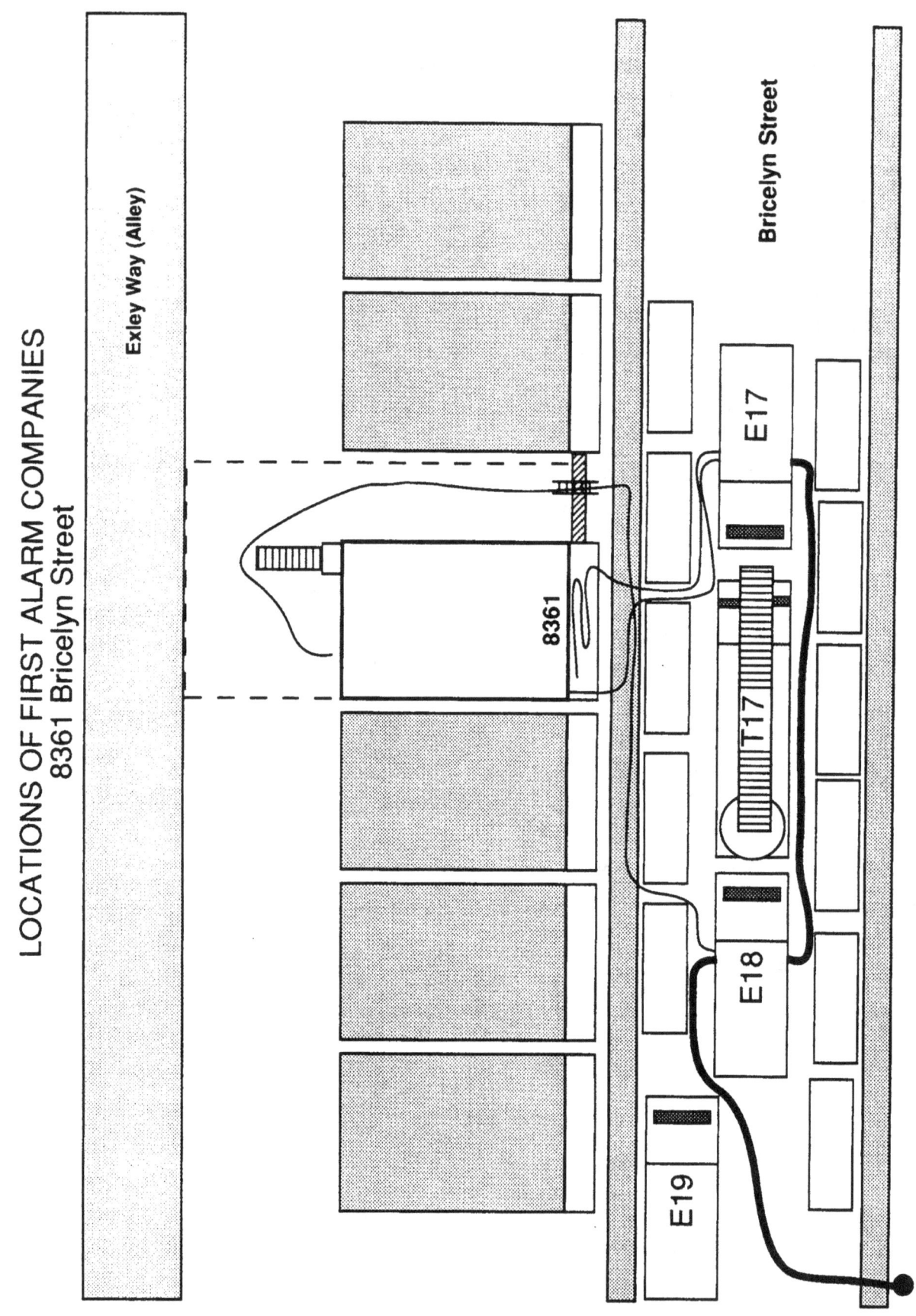

LOCATIONS OF FIRST ALARM COMPANIES
8361 Bricelyn Street

Firefighters in Distress

The captain came down from the roof, donned an SCBA, picked up an axe, and entered the house through the front door. He followed the charged hoseline down the hallway then made the right turn down the stairs. As he descended toward the family room he felt the stairs give way under his weight; the fire had weakened the stair supports. He fell approximately 10 feet and landed on the basement stairs.

He had fallen into the basement, close to the seat of the fire. Although he could see the flames and feel the heat, there was spray from a hose stream that provided partial protection. He was able to climb back up the stairs to the family room level where he encountered the captain of Engine 17 who was operating the hoseline. Although the smoke was too heavy to see clearly, he later stated that he was aware of the presence of all three crew members from Engine 17 in the immediate area.

It is not clear what happened during the next few minutes. At some point all three personnel from Engine 17 realized they were running short of air and needed to leave, but could not find a way out of the family room. They were unable to find a window or any other exit and exhausted their air supplies. At some point the hoseline, which was suspended over the breach in the stairs, burned through.

At least one of the firefighters is believed to have removed or loosened his facepiece and some attempts were made to share the air that was available by buddy breathing. All three crew members from Engine 17 were rendered unconscious due to carbon monoxide inhalation and/or oxygen deficiency. The captain from Truck 17 was also incapacitated by carbon monoxide and was down on his hands and knees, almost unconscious.

First Rescue

Engine 18's line had been successful in knocking down most of the visible fire in the basement, and the crew was preparing to advance into the basement through the rear door to complete extinguishment. At approximately 0056 hours a firefighter from Engine 18 broke one of the family room windows to provide additional ventilation. When the window was broken he heard the taps from the low pressure alarm bell on an SCBA and a moan. He entered through the window and discovered the semi-conscious captain down on the floor, close to the window.

Assisted by another firefighter and the EMS personnel from Rescue 1, he was able to pull the captain out the window and lower him to the ground. The first radio report of a firefighter down was transmitted by Battalion 3 at 0058 hours. At 0101 Battalion 3 reported "we found him, he's all right."

Second and Third Rescues

As the EMS personnel began to treat the captain, who was barely conscious, he gasped that there were more firefighters inside and he thought they were dead. Several firefighters returned to the side of the house where they encountered two firefighters from Engine 19 who had just crawled through the same window.

The crew from Engine 19 had been stopped at the front door unable to advance for several minutes due to the heavy heat and smoke in the front hallway. It appears that the heat and smoke conditions at that time were caused by the breach in the stairs, where the captain had fallen through to the basement. The captain and one of the firefighters depleted their air supplies and returned to their apparatus to change air cylinders, leaving one firefighter on the porch with the hoseline.

While the captain and firefighter were at the apparatus changing cylinders, the driver of Engine 19 donned an SCBA and proceeded to the front door, where he encountered the other firefighter from Engine 19. By this time the interior heat and smoke conditions were significantly reduced, possibly because the hoseline had ruptured directly over the opening in the stairway. The driver entered the house, following Engine 17's hoseline, made the turn down the stairs, and fell into the same hole where the captain from Truck 17 had fallen through the stairs.

The firefighter followed the driver and also fell through the hole in the stairway. They assisted each other back up the stairs to the family room, then walked across the room toward a light that was shining through the broken window. They were assisted from the window by the firefighters outside and turned over to the EMS personnel in the back yard for treatment.

There was a period of confusion after the two additional firefighters were removed from the window. The first report of additional missing firefighters was transmitted on the EMS radio channel at 0106 hours. At 0107 there was a transmission reporting that, "17 Engine Captain reports three men in; three men out." This was followed two minutes later by a report of, "three firefighters down and one still missing", also on the EMS channel.

The EMS supervisor and the acting deputy fire chief, both of whom had just arrived, went to the rear yard and confirmed that three personnel had been rescued from the house and were being treated. The acting deputy chief then requested a third alarm, because several of the firefighters on the scene were now assisting in treatment of the three rescued personnel. The third alarm was dispatched at 0110 hours.

The acting deputy chief returned to the front of the dwelling where he met the acting battalion chief and assumed command of the incident. The two command officers believed that all of the missing firefighters were out of the house. At 0116 hours the EMS supervisor reported on the EMS radio channel that all firefighters had been accounted for.

By this time the fire had extended into the attic and was breaking out in the third floor bedrooms. The second and third alarm companies were assigned to fire control efforts, while the three injured firefighters were transported to hospitals in two ambulances.

Missing Firefighter Located

Engine 7, a second alarm company, advanced a 1 3/4 inch line to the rear porch and attacked the fire that had broken out in the kitchen, while Engine 18's line was advanced into the basement. The fire had extended into the cabinets under the kitchen counter tops and weakened the floor joists, causing the kitchen floor to sag at the center of the room. It took a considerable amount of time for Engine 7 to gain control of the situation in the kitchen and then advance the line into the family room. By his time the interior smoke condition had been relieved and visibility had improved.

Advancing into the family room, the crew from Engine 7 found the three unconscious personnel from Engine 17. (See diagram on the following page.) At 0139 hours the Incident Commander was informed that three additional firefighters had been found and were being removed from the building. Additional personnel responded to the urgent requests for assistance and the three firefighters were removed to the rear yard.

The EMS personnel determined that all three were in cardiac arrest and initiated resuscitation efforts.

The assistant fire chief, who had responded on the third alarm and was in the process of assuming command of the incident, requested two additional alarms to provide personnel for fire suppression, since most of the personnel on the scene were involved in the efforts to rescue and revive the three firefighters. The three crew members from Engine 17 were all dead on arrival at hospitals.

The fire was controlled without further incidents. The chronology for the incident is detailed on the following below.

CHRONOLOGY

00:22:00	Call to 9-1-1 from 8366 Bricelyn Street reporting structure fire.
00:22:53	First Alarm Dispatch, Zone 3-15: E17, E18, E19, T17, 4042.
00:27:42	Engine 17 on scene, reports "smoke from 1st floor."
00:28:09	Battalion Chief 3 (4043) responding in place of Battalion Chief 2 (4042).
00:28:09	Truck 17 on scene.
00:32:27	Engine 18 on scene.
00:32:48	Engine 19 on scene.
00:37:04	Battalion Chief 3 on scene.
00:38:38	Battalion Chief 3 reports, "3 story…heavy smoke from 2nd and 1st floor."
00:41:50	Medic 11 (5111 on scene. (EMS Channel).
00:46:20	Battalion Chief 3 requests 2nd alarm to respond to the rear.
00:46:55	2nd alarm dispatched: E15, E7, T8, DC4, 4539.
00:49:49	Rescue 1 (5101) on scene.
00:54:30	Engine 15 on scene.
00:54:42	Truck 8 on scene.
00:55:36	MAC2 (Air Unit) on scene.
00:56:54	Engine 7 on scene at Exley Way.
00:58:27	Battalion Chief 3 reports "Fireman down inside, trying to get him out".
00:59:40	Deputy Chief (404) on scene.
01:00:54	EMS supervisor (502) on scene.
01:01:19	Battalion Chief 3 reports "We found him. He's all right."
01:05:08	Unit 4539 on scene.
01:05:58	Medic 11 reports "We have more firemen in the building." (EMS Channel)
01:07:58	Medic 11 reports "Fireman out of the building." (EMS Channel).

INTERIOR DETAILS - 8361 Bricelyn Street

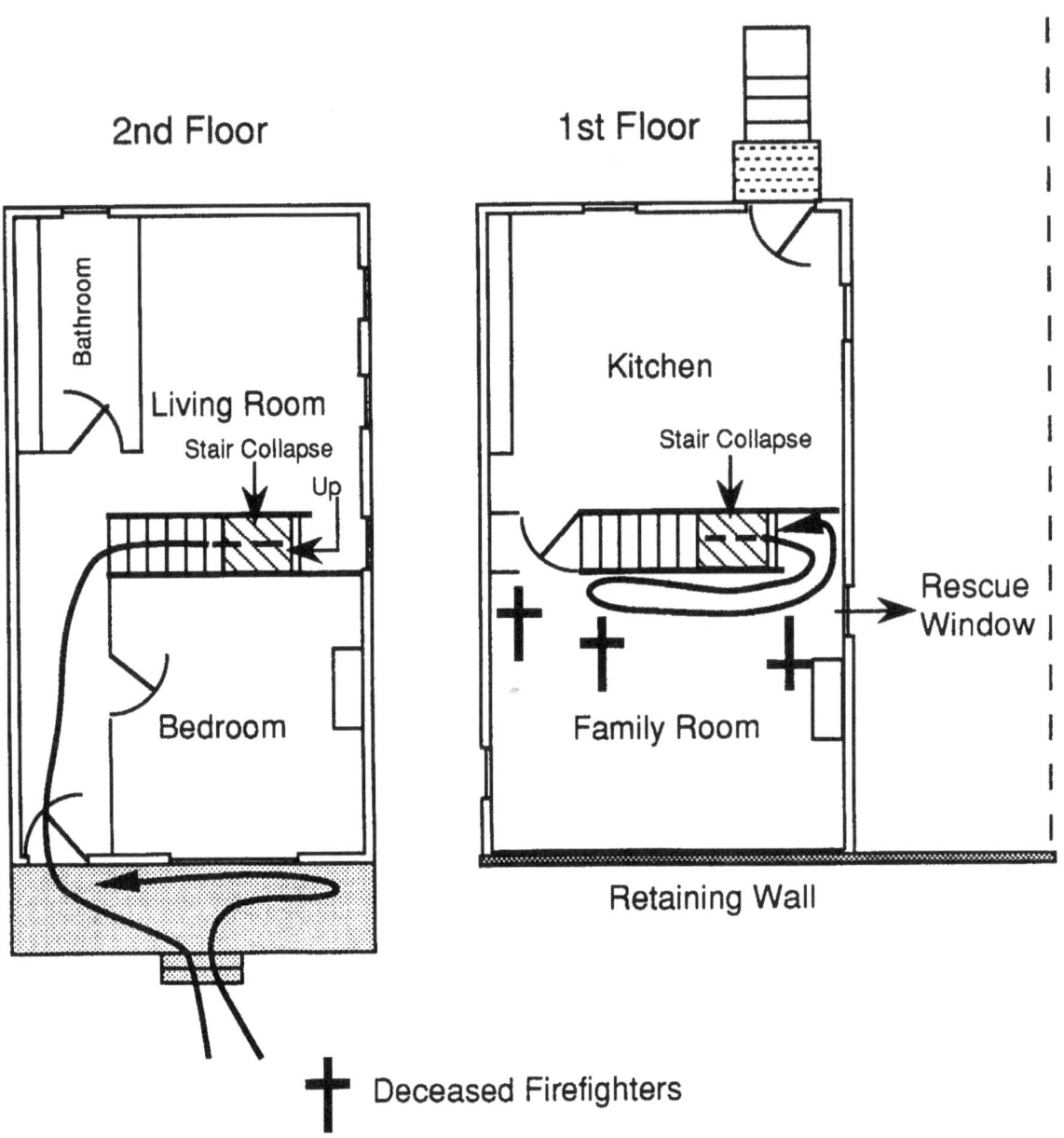

2nd Floor

1st Floor

Bathroom

Living Room

Stair Collapse

Up

Bedroom

Kitchen

Stair Collapse

Rescue Window

Family Room

Retaining Wall

✝ Deceased Firefighters

01:09:45   Dispatcher asks Incident Commander: "Is there a report of a fireman missing and two others down?"

01:09:50   Deputy chief reports "The medics are back here. We got them all."

01:10:06   3rd alarm dispatched: E11, E8

01:11:53   Battalion Chief 3 asks deputy chief "Do we have everybody we started with?"

01:14:04   Deputy Chief reports he is going out front.

01:15:29   Engine 11 on scene.

01:15:48   EMS Supervisor reports "All firefighters are accounted-for." (EMS Channel)

01:20:35   Engine 8 on scene.

01:27:01   EMS Supervisor reports "All three injured firefighters transported." (EMS Channel).

01:39:43   Battalion Chief 3 reports "Three firemen down. They are trying to get them out of the side."

01:41:48   Assistant Fire Chief (41) on the scene.

01:43:24   EMS Supervisor reports "Three firefighters in cardiac arrest." (EMS Channel).

01:46:57   Deputy Chief requests 4th alarm.

01:51:21   Assistant Fire Chief assumes command.

02:04:36   EMS Supervisor reports "Three firefighters have been removed from side 3." (EMS Channel).

02:09:54   Assistant Fire Chief requests 5th alarm.

02:43:34   Assistant Fire Chief declare fire under control.

ANALYSIS

This incident demonstrates how easily firefighter fatalities can occur when simple and basic operational problems occur together in the worst possible sequence. There was no indication that this incident was more than a "routine" dwelling fire and many veteran members of the Pittsburgh Fire Bureau expressed disbelief that three firefighters could have been "lost" for more than an hour or could have died in a small room that had two windows to the exterior, within six feet of the ground level.

The combination of circumstances that resulted in these three deaths illustrates that every fire situation presents inherent dangers. There is no single problem event or circumstance that can be blamed for what happened. Several contributing factors can be identified that combined to turn a seemingly routine fire into a tragedy of immense proportions. The detailed analysis of the incident points out several areas where changes should be made, particularly in the areas of training and incident management.

Known and Unknown Factors

All three firefighters died from asphyxiation, which involved both carbon monoxide inhalation and oxygen deficiency. This occurred as a direct result of exhausting their air supplies and being unable to find an exit from the family room. It is assumed, but it cannot be confirmed, that all three were using their SCBAs, with facepieces in place, for the entire time they were inside the house until their air supplies were depleted.

It is known that the three crew members from Engine 17 exhausted their air supplies and that their vision was fully obscured by heavy smoke, at least during the period when they were running out of air. The fire did not involve the room where they were found; after the fire there was only moderate heat damage at the upper levels of the room and their protective clothing and equipment showed no signs of damage or deterioration caused by exposure to the fire.

It is not known if the deceased firefighters knew immediately that the fire was in the basement, which was two levels below their point of entry, or if they spent some time looking for the fire on the first or second floors. There was no indication that they had attempted to advance the hoseline down into the basement or into the kitchen.

It is not known if the hoseline that was operated through the rear window into the basement caused an opposing hoseline situation that may have pushed additional smoke and heat toward the three deceased firefighters as they tried to advance down the basement stairs.

Their return path back up the stairs was blocked by the fire coming up through the collapsed section. Their hoseline, which was suspended over the collapses stair section, eventually burned through and ruptured. The hoseline was found with the nozzle open at the bottom of the stairs, which suggest that they were trying to control the fire that was coming up through the opening when the hose ruptured. It appears that the firefighters could have escaped through the kitchen or through the windows in the family room. With the door to the basement stairs in the open position their path into the kitchen would have been blocked and the smoke and heat coming up the stairs would have presented an additional obstacle. If they had closed the door at the top of the stairs they might have been able to walk or crawl out through the kitchen, if they had known about the kitchen door and windows.

The trapped firefighters did not appear to have located either of the two windows in the family room. The insides of the windows were covered by the plastic panels, which left a smooth interior surface that would have made them somewhat more difficult to locate than regular windows. There is no indication that they were searching for windows by feeling their way along the walls and no indication that any attempt had been made to break either of the windows from the inside. There was at least one axe in the room, which could have been used to break out a window.

The exact sequence of events inside the family room remains a mystery and will probably never be known. The following analysis examines several different issues that relate to the three victims and their equipment, in an attempt to better understand what appears to have happened to them. It also examines the overall management of the incident and how that is related to the outcome.

SAFETY EQUIPMENT

Self-Contained Breathing Apparatus

The SCBA units worn by the three fatally injured firefighters and the unit worn by the surviving captain were impounded at the scene for examination and testing. The SCBAs used by the deceased members were found with the air supplies fully exhausted, while the SCBA worn by the survivor still had 300 psi of air pressure remaining in the storage cylinder.

Autopsy Results – All three firefighters died from asphyxiation. Two of the three victims had high carboxyhemoglobin levels, in the 40 to 50 percent range, which indicates that they died primarily from smoke inhalation. The third deceased firefighter had a relatively low carboxyhemoglobin level, approximately 10 percent, and appears to have died primarily from hypoxia (oxygen deficiency). The captain from Truck 17 also had a carboxyhemoglobin level in the 40 to 50 percent range, which was incapacitating and potentially fatal.

Observations – Two of the three deceased firefighters were reported to have been found with their facepieces in place and their SCBAs in normal operating configuration. It could not be confirmed if the third firefighter's facepiece was in place and secure when he was found, or if the breathing tube had been disconnected from the regulator. Only one of the deceased firefighters was confirmed to have the facepiece still tightly in place when found.

It is not known if all three victims used their SCBAs for the entire time they were inside the dwelling and exposed to the products of combustion. All three entered the dwelling with their SCBAs on their backs, but it is not known if their facepieces were in place at the time they entered. It is possible that one or more of the individuals may have delayed using their SCBA and inhaled significant quantities of carbon monoxide before realizing that respiratory protection was needed.

The crew of Engine 17 may have spent some time looking for the fire on the first floor level, thinking they were in the basement, before they located the basement stairs and realized that the fire was below them. They may have suddenly encountered a heavy concentration of smoke and heat when they opened the door at the top of the stairs and only donned their facepieces at that time.

Test Results – The four SCBAs were examined and tested by the Division of Safety Research of the National Institutes of Occupational Safety and Health (NIOSH) at their facilities in Morgantown, West Virginia. Three series of tests were performed on the four units by NIOSH.

In the first series of tests, each of the four units was found to be deficient in one or more of the performance tests that are required by NIOSH for certification of an SCBA. The tests indicated that three units failed to meet the required 200 liter per minute air flow rate, two failed to meet the 30 minute service duration requirement, and two failed to maintain positive pressure throughout the respiration cycle. The low pressure alarms on all four units were set to activate below the required level and one low pressure alarm intermittently failed to operate.

The breathing resistance test could only be performed on two units and one of those units exceeded the static pressure limit of 1.5 inches of water column inside the facepiece. One facepiece could not be tested because of the contaminants inside the facepiece and the exhalation valve. Another facepiece could not be properly tested because the exhalation valve spring was dislodged and the valve would not seal in the closed position.

While problems were noted with each of the four units, all four were found to be functional and capable of delivering air to the users. Although three of the four units failed to meet the 200 liter per minute flow rate, they should have been able to deliver an adequate flow to meet the user's demand rates, unless the users were engaged in extremely strenuous activities. It was also determined that all four SCBAs could maintain a positive pressure while in use, except for some potential inward leakage during inhalation on two of the units. The leakage tended to let air escape from the facepieces more than allowing outside air to enter, which is the objective of positive pressure. The low pressure alarm settings were only slightly below the specified levels.

A second set of tests was conducted by NIOSH to evaluate the amount of inward leakage that could have occurred with the four SCBAs. These tests indicated that there may have been some inward leakage in two of the units, caused by a combination of an inadequate air flow rate into the facepiece and leakage through the exhalation valves or around the facepiece seals. The rate of inward leakage was measured for each unit and the amount of carbon monoxide exposure to the users was estimated, using calculations that were based on the concentration of carbon monoxide that can be anticipated inside a burning dwelling and the time the users would have been exposed.

These tests indicated that the leakage would not account for the high carboxyhemoglobin levels that were found in three of the victims. The victim who had a low carboxyhemoglobin level was using one of the units that was found to be leaking and the estimated rate of CO leakage with this unit came closer to being consistent with the autopsy findings. Even in this case, however, the calculations based on the measured leakage rate did not fully account for the carboxyhemoglobin levels that were reported in the autopsy results.

A third series of tests was conducted to determine the cause or causes of the performance deficiencies that had been observed in the first series of tests. The regulators were examined and initial pressure and flow reading were taken. Standard maintenance and calibration procedures were then performed to bring them into compliance with the manufacturer's specifications. It was determined that all four regulators could be brought into compliance through the performance of the routine maintenance and calibration procedures on the regulators and low pressure alarms. When retested all four units met or exceeded the NIOSH performance requirements and the manufacturer's standards.

These findings underline the need for regular inspection and maintenance programs for all SCBA units, with the work performed by properly trained technicians, following the manufacturer's recommended procedures. Some of the problems that were found may have resulted from untrained members attempting to perform maintenance procedures on their SCBAs. The problems with exhalation valves and facepiece seals should have been detectable by the users through daily user checks or by testing for a tight seal when donning the facepiece.

When the initial reading were taken for the third test series, the restricted air flow problems that had been observed in the first series did not reoccur. This suggests that the initial problems may have been attributable to moisture or debris which got into the facepieces and regulators at the scene of the fire and then evaporated or were released with time and repeated testing. It is not known if all of the performance deficiencies that were observed in the initial tests were consistent with the condition of the units when they were used at the Bricelyn Street fire.

The four units had been removed from the house and were outside in sub-freezing weather when they were impounded for the investigation. They were stored in a secure location, but were not closely inspected until several days after the fire and the regulators were not opened until two months after the incident. It is possible that a combination of moisture and foreign matter may have gotten into the regulators before they were impounded.

Relationship to Autopsy Results – Based on the test results and autopsy findings, it does not appear that the performance deficiencies that were found in the SCBAs were the primary causal factors in the deaths of the three firefighters. The fact that the three victims exhausted their air supplies appears to be the primary causal factor.

The high level carboxyhemoglobin levels, which are consistent with unconsciousness, coma or death, depending on the individual and the length of exposure, suggest that three of the four individuals were exposed to the fire atmosphere for some time without the protection of their breathing apparatus. It cannot be determined if this occurred after their air supplies were exhausted, during attempts to "buddy breathe" or during periods when the firefighters may have been working without the protection of their SCBAs

It is not uncommon for firefighters to remove or loosen their facepieces after running out of air and the reported autopsy results are similar to the findings from previous situations where this has occurred. High level exposure to carbon monoxide is known to quickly cause disorientation and compromise motor skills, which would have made it more difficult for the firefighters to find a means of escape from the dwelling. The elevated temperatures inside the house may have also contributed to their rapid incapacitation and disorientation.

At least one of the victims may have removed or loosened his facepiece and one or two of the others may have attempted to "buddy breathe" with him. Experience with other cases of attempted buddy breathing indicates that both individuals may be very quickly exposed to high concentrations of CO or other contaminants, unless the SCBAs are equipped with special fittings to allow air to be shared. This may quickly result in two individuals in distress, instead of one being capable of helping to rescue the other.

The most likely explanation for the low carboxyhemoglobin level found in the third deceased firefighter is the leakage through the exhalation valve or around the facepiece seal. This victim was found with the facepiece still tightly in place. This amount of exposure to carbon monoxide would be likely to cause a headache and minor impairment to the firefighter, but it is not likely to result in disorientation or significant impairment of motor skills.

PASS Devices

All four of the SCBAs had personal alert safety system (PASS) devices attached to the harness traps and all four were found in the off position. This indicates that the firefighters did not turn on their PASS devices before entering the building and did not manually activate them when they found themselves in trouble. If the PASS devices had been operating there is a high probability that they would have been heard by other firefighters who were in the room or at the window and a reasonable expectation that the victims could have been successfully rescued.

The PASS units were tested and found to be operational, however, the alarm signal sound outputs for all four units were below the levels specified in the 1988 edition of the NFPA Standard 1982, which was in effect when they were purchased. Even with the reduced sound output levels, it appears likely that they would have been heard, especially with four units operating in a small room.

Portable Radios

Two of the victims had portable radios; however, both radios were found in turnout coat pockets and there was no indication that the victims had attempted to use them to call for assistance. No transmissions were received from either radio during the time that the firefighters are believed to have been in distress. The only radio transmission that was recorded from either radio was the initial report given by the captain of Engine 17 before entering the structure.

The firefighter's radio was operational when it was tested later than night. The captain's radio, which had been used to transmit the initial report, had a dead battery and a blown fuse when it was tested, but functioned properly when these items were replaced. Both of these problems may have resulted from exposure of the radio to water, which may have occurred before or after the victims were removed from the house.

It was not determined if the radios were set on the proper channel for the incident.

INCIDENT MANAGEMENT

The Pittsburgh Fire Bureau has adopted the Incident Command System as a standard operating procedure. However, the degree of application of the system depends on the Incident Commander.

The Incident Command System and the Pittsburgh Bureau's SOP #6 stress the importance of an early assumption of command so that all operations, from the outset, are directed by one individual and based on a single attack plan. The procedures call for an early arriving company officer to assume command of the incident if there is no chief officer at the scene. The early assumption of command and development of a strategic plan are particularly important considerations to ensure that operations are conducted safely.

None of the company officers assumed command of this incident and there is no indication that anyone was attempting to direct or coordinate the actions of the first alarm companies, except for the request to have the next arriving engine company provide a supply line for Engine 17. All of the first alarm companies committed themselves to various operations based on their own initiatives. The first officer to officially assume command of the incident was the acting battalion chief, who arrived 10 minutes after the first companies.

The acting battalion chief announced that he was establishing command and positioned himself at the front of the dwelling. He did not receive any initial information on interior conditions or activities and did not receive any progress reports from the operating companies over the radio. His knowledge of the situation was limited to the operations he could observe from the front of the dwelling.

The acting battalion chief was placed at a significant disadvantage due to his late arrival at the fire scene and the complicated arrangement of the dwelling's interior. He did not know that there was an intermediate level between the front entrance and the basement. He had no information on the specific location and actions of Engine 17 or the fire conditions they were encountering. From the front he had very limited view of the structure to make an exterior size-up.

The acting battalion chief did not give any directions or assignments to the first alarm companies over the radio. The only specific instructions that were recorded on the radio tape were to direct the second alarm companies to come in from the rear. Face-to-face communications would not have been recorded.

The acting deputy chief arrived while the first rescues were being made. He initially went to the rear yard to size-up the fire situation, where he encountered the EMS personnel treating injured firefighters. After spending several minutes dealing with that situation, he completed the exterior size-up and went to the front where he assumed command of the incident from the acting battalion chief. He did not have information on interior fire conditions or on the locations, assignments and accountability for the interior crews when command was transferred.

The acting deputy chief had an initial concern about accounting for all personnel at the scene, but he received information that all personnel had been accounted for. This relieved his concern and caused him to focus on fire suppression efforts until he was informed that the three personnel from Engine 17 had been found unconscious inside the dwelling.

EMS Coordination

The EMS units that respond to fires are not under the direction of the Fire Bureau Incident Commander. The EMS units communicate with each other and with their dispatcher on an EMS radio channel, but they have to coordinate face-to-face with Fire Bureau units or go through their dispatcher who is located in the same Communications Center as the fire dispatcher. An EMS supervisor responds on the second alarm to establish a liaison at the command post.

The two EMS rescue units in Pittsburgh carry protective clothing and breathing apparatus so that the crews can enter building to assist with rescue operations and the removal of injured patients, however, they are not integrated into the Fire Bureau's incident management structure. The separate radio channels make it difficult to coordinate operations between fire and EMS units that may be working side by side. The lack of a unified command system and accountability for all personnel in the fire area can create potentially dangerous situations.

At this incident there was a parallel flow of communications on the fire and EMS radio channels. Significant information relating to the identities of the individuals who had been rescued was not effectively communicated back to the Incident Commander. The rescued firefighters were being treated by EMS personnel and at one point the EMS radio channel quoted "17 Engine captain reports three men in – three men out." (This apparently came from an EMS employee who recognized the captain who was being treated and associated him with his previous assignment at Engine 17). A later report over the EMS radio advised that all personnel had been accounted for. These messages added to the confusion over who had been rescued and who was missing.

The EMS supervisor, who responded on the second alarm, initially went to the rear yard where the firefighters were being treated. After the patients had been transported, he went to the front of the house and established the EMS liaison at the Command Post. The identities of the individuals who had been transported were not provided to the Incident Commander.

SAFETY AND ACCOUNTABILITY

Many fire departments that routinely use command and control systems to keep track of the locations and assignments of companies have added accountability systems to keep track of the individuals assigned to each company. This information is used during the course of operations to account for all personnel at regular intervals and verify that they are safe.

An accountability check involves communicating with each company officer to ensure that they can personally account for each assigned individual. Accountability checks are required at regular intervals, usually every 15 to 20 minutes. A significant event, such as a report that firefighters are injured or missing, or a change in the strategic plan, calls for a special accountability check.

The Pittsburgh Fire Bureau had a written procedure for a safety identification system to be established on multiple alarm incidents; however, its implementation was at the discretion of the Incident Commander and it did not require regular accountability checks. Name tags are issued to all personnel and deposited on the apparatus when the individual is assigned to that crew. When the system is implemented, all of the tags are picked up and assembled on a control board which is used to track the location, status, and assignment of each company at the incident. This system provides a crew roster for the operating companies at the command post.

On all multiple alarms Engine Company 39 is placed out of service and the crew responds with Unit 4539 to perform safety-related functions. The company officer becomes the Incident Safety Officer and two for the crew members are responsible for setting up the control board at the command post. The remaining crew member is assigned to refill SCBA cylinders.

Unit 4539 arrived with the second alarm companies and the personnel assumed their pre-assigned roles. The control board was set up, but the name tag system was not implemented. Without the personnel identification system there was no roster to verify the assignments of the personnel who had been rescued and ensure that no one else was missing.

If an accountability check had been required 20 minutes after arrival, the time would have coincided with the period when it appears that the crew of Engine 17 was running out of air and realizing they were in trouble. A rescue operation initiated at that time might have found and rescued them before they were fatally overcome. An accountability check after the first three members had been rescued also would have revealed the crew of Engine 17 was missing.

Crew Integrity

One of the important considerations in all incident management systems and accountability systems is crew integrity. All members of a company should function as a team, taking their direction from the company officer. The company officer is expected to be accountable for the position, function, and safety of all members of the company at all times. Accountability systems require company officers to verify at regular intervals that they can account for all company members.

There are situations where it is appropriate to split a crew into smaller teams. However, each team that is operating in a fire area should have a minimum of two members and at least one of those members should have a portable radio to maintain contact with the company officer or with the appropriate supervisor in the incident command system.

Three of the four companies on the first alarm failed to maintain crew integrity. The three members from Engine 17 were the only crew that stayed together. The three individuals who were rescued were not in contact with their company officers and their company officers did not recognize that members of their crews were missing until after the three personnel from Engine 17 had been found. The confusion over the identities of the individuals who had been rescued contributed significantly to the failure to recognize that Engine 17's crew was missing.

TEMPORARY ASSIGNMENTS

The fact that 12 out of 17 personnel on the first alarm were in out-of-class assignments or filling positions for other individuals added to the confusion at this incident. The officers and crew members were not used to working together and their helmet markings did not indicate their temporary assignments. This reinforces the need for an accountability system, including a means to visually indicate temporary assignments.

Several individuals were working in higher level positions because regularly assigned officers were absent or "moved up" that night. The initial incident commander was an action battalion chief who normally functions as a company officer and was not used to managing incidents at this level. One of the individuals who was rescued was a captain from an engine company, who was working an extra shift in a firefighter's position on a ladder company, reporting to an acting lieutenant. Some of these individuals may have been unsure of their responsibilities or had difficulties with reporting relationships.

EMERGENCY SURVIVAL ACTIONS

While the exact sequence of events will never be known, it is evident that the three crew members from Engine 17, at some point in time, realized they were in trouble. Assuming that they entered the structure with full air cylinders and consumed the air at a normal rate, they probably would have been running short of air between 0045 and 0050 hours, which is close to the time the captain from Truck 17 encountered them.

The captain later reported that none of the crew members from Engine 17 seemed to be in distress at the time he first encountered them, but they began to have critical problems within a short time after he arrived in the family room. He described a sequence of events under zero visibility conditions in which low pressure alarms were sounding and attempts were being made to share the remaining air supply by "buddy breathing."

The actions of the victims may have been impaired by the combined effects of carbon monoxide inhalation and heat stress, which are likely to cause confusion and interfere with motor skills. It is not known what efforts they were making to find a way out of the house; however there is no evidence that they located the windows or the doorway leading to the kitchen.

It is not known why the trapped personnel did not use their portable radios to call for assistance when they realized they were in trouble or manually activate their PASS devices to sound an audible alarm. The analysis of several incidents involving firefighters in critical situations also suggests that firefighters are often reluctant to admit they are in trouble until it is too late to effectively call for assistance.

Firefighter training should include standard survival actions that should be followed in emergency situations. This type of training, which is required by NFPA Standard 1500, should include an emphasis on early recognition of being in trouble and calling for assistance without delay. A standard procedure, such as a 'MAYDAY" over the radio, should trigger a priority response to locate and assist the personnel who are in trouble.

As noted in the section on breathing apparatus, experience from similar incidents also suggests that "buddy breathing" should be discouraged, unless the breathing apparatus has been provided with special connections to interconnect units. In most situations it appears to be more effective for members who still have air in their cylinders to use it to rescue those in trouble or to summon assistance than to attempt to share the limited available supply. All members should be well trained in the procedures that should be followed if an SCBA malfunctions or the air supply is depleted.

Training should emphasize the need to identify alternate exists in all situations. Company officers should always be aware of the environment where their companies are working, keeping an eye on conditions, looking for alternative exits, and thinking about how to react if conditions change suddenly. This requires company officers to resist the temptation to become too involved in performing manual tasks to fulfill the responsibilities of a supervisor.

RAPID INTERVENTION TEAMS

The 1992 edition of NFPA 1500 also requires the assignment of designated "rapid intervention teams" at working incidents. The mission of a rapid intervention team is to be ready to take immediate action to rescue or provide back-up for firefighters in trouble or to search for companies that fail to respond to a request for an accountability check. This is recommended to be a specifically designated crew or company, assigned to stand-by, wearing full protective clothing with SCBA ready for use.

LESSONS LEARNED AND REINFORCED

1. Incident Command procedures should be utilized at all multi-company incidents. The early establishment of incident command is essential to initiate operations in an organized and coordinated manner. All personnel should be trained in the basic concepts of the incident command system and fully capable of performing their normal functions, as well as any other functions they could be assigned to perform. Command and staff personnel must be assigned to create and operate an organization that is in proportion to the incident. All units that operate in the fire area, including EMS units, should be integrated into the incident command system.

2. A qualified Incident Safety Officer should be assigned to all working incidents to ensure that all safety concerns are fully addressed.

3. Accountability procedures should be implemented at all working incidents. A standard approach to accountability is needed to routinely verify that all personnel are safe at regular intervals and to specifically account for all personnel when there is a change in strategy or when an unusual event occurs at an incident.

4. A Rapid Intervention Team should be assigned to stand-by at all working incidents to provide assistance to companies or individual personnel who are in danger.

5. Crew integrity should be maintained. Companies should work together as teams, under the direct supervision of their company officers. The company officer should be in contact with all assigned personnel at all times. If companies are split into teams, each team should have a separate identity and a portable radio and should be tracked as a separate unit in the accountability system.

6 Emergency survival procedures should be a standard component of firefighting training. All personnel should be well trained in the actions that should be taken in emergency situations, such as becoming lost and disoriented or running out of air in a hazardous area. This training should include when and how to request assistance and a standard approach to be implemented when a report of personnel missing or in distress is received.

7. A Self-Contained Breathing Apparatus maintenance program is essential. All SCBA units should be inspected and function tested at every shift change and after each use. Any units with defects should be immediately taken out of service for inspection, evaluation, and repair by qualified technicians. Untrained personnel should not be permitted to perform any procedures that involve disassembly of the SCBA regulator.

All SCBA units should be regularly inspected and should have preventive maintenance procedures, including calibration and testing, performed by qualified technicians following the manufacturer's documented recommendations.

8. Personal Alert Safety Systems should be utilized at all incidents where firefighters operate in hazardous areas. Every SCBA should be equipped with a functioning PASS device. Members should be trained to always turn on the PASS when working in a potentially hazardous area and should manually activate the PASS alarm if they are in danger. The PASS devices should be function tested daily and regularly inspected and maintained by trained personnel. Technological advances in PASS devices should be researched to determine if more advanced and reliable units can be obtained.

9. Hillside structures require special attention. Fires in hillside structures should focus special attention on identifying the different levels and which levels have exterior access from different sides. It is particularly important to avoid placing companies over a fire when they are not aware that the fire is below them.

APPENDIX A

Bureau Initiatives

Following the completion and presentation of a report on this incident by an internal Board of Inquiry, the following actions were implemented by the Pittsburgh Fire Bureau.

Additional changes were being considered at the time this report was completed.

BUREAU INITIATIVES

October 5, 1995

Operations

C   Dispatch transmit to Command a time announcement every 15 minutes (elapsed time of fire)

C   2nd battalion chief respond as safety officer to all structure fires

C   Safety unit/command board (Engine 39) respond to all structure fires

C   Standby company (Go-Team) dispatched 2 alarm structure fires or greater

C   Establishment of Hazard Zones at emergency incidents

C   Established a first alarm water supply policy (forward lay)

C   Expanded requirement of first arriving company officer radio status report

C   Expanded requirement of first arriving company officer radio status report

C   Established trigger word to immediately clear fireground channel for emergency traffic.

C   Established policy of mandatory activation of PASS Device in a hazard zone.

C   Reinforce use of existing Incident Command and accountability systems

C   EMS Bureau units reassigned to fireground channel during structure fires

C   Established policy that company officers will not operate nozzles on fire attack lines

Equipment

P   Upgrading existing PASS Devices

P   Relocate fireground channel(s) positions on portable radios

P   Upgrade SCBA apparatus to state of the art

C   Revised and expanded S.O.P. on SCBA concerning inspection/use/care

C   Increased maintenance/inspection cycle on SCBA

Training

CAll chief officers attend ICS/Tactics and Strategy training course

PCompany officers attend ICS/Tactics and Strategy training course

PProvide in-service SCBA training program

PProvide in-service training on firefighter survival techniques

C = Completed   P = Planned next two quarters

APPENDIX B

Photographs of Fire Scene

1. Front view of 8361 Bricelyn Street.

2. Rear view of 8361 Bricelyn Street.

3. Side yard showing retaining wall and window where firefighters were rescued (above lawn chair).

4. Window in corner of family room showing plastic panel which has partially melted and bent-over due to heat.

5. Inside view of window where firefighters were rescued.

6. View from family room looking toward kitchen. Partially open door is at the top of the basement stairs.

7. View of corner of the family room at top of basement stairs.

www.ingramcontent.com/pod-product-compliance
Lightning Source LLC
Chambersburg PA
CBHW081408170526
45166CB00010B/3255